HAM RADIO FOR BEGINNERS:

The Ultimate Manual Guide to Build and Operate Your Amateur Radio as a Newbie

BY

ROBERT SHOCKLEY

TABLE OF CONTENTS

INTRODUCTION

In a world dominated by advanced communication technologies, ham radio, also known as amateur radio, stands out as a timeless and resilient means of communication. Ham radio is a fascinating hobby that empowers individuals to explore the vast realms of radio frequency communication, allowing them to connect with people from around the globe, experiment with cutting-edge technologies, and contribute to emergency communication networks. This captivating and dynamic hobby has been a source of innovation, camaraderie, and public service for decades.

At its core, ham radio is a non-commercial, two-way radio communication service that is open to individuals of all ages and backgrounds.

Ham radio operators, commonly referred to as "hams," use designated radio frequencies to establish contact with fellow enthusiasts, exchange information, and engage in a variety of activities such as contesting, chasing rare radio signals, and participating in community events. The allure of ham radio lies not only in its technical aspects but also in the sense of community that binds together operators across the globe.

One of the distinguishing features of ham radio is the unique call sign assigned to each operator, serving as a personal identifier on the airwaves. These call signs are not only functional but also carry a sense of tradition and identity within the ham radio community. Whether engaging in casual conversations with other hams, participating in organized events, or providing vital communication during emergencies, the call sign becomes a badge of honor for operators as they navigate the vast spectrum of possibilities within the hobby.

The history of ham radio is rich and deeply intertwined with the development of radio communication itself. From its early roots in the late 19th century to its evolution as a popular hobby in the 20th century, ham radio has played a pivotal role in technological advancements and has been a source of innovation in communication. Today, it continues to thrive as a diverse and inclusive community, attracting individuals with interests ranging from electronics and engineering to international friendship and public service.

For beginners embarking on their ham radio journey, the initial steps may seem daunting, with a myriad of technical terms and regulatory requirements. However, the entry into ham radio is facilitated by a supportive community of experienced operators, readily available educational resources, and licensing procedures that ensure a basic understanding of radio operation and regulations. Aspiring hams will find that the learning process is not only educational but also deeply rewarding, opening doors to new friendships, technical skills, and a sense of belonging within the global ham radio community.

In this guide for beginners, we will delve into the fundamental aspects of ham radio, exploring the basics of radio communication, the licensing process, essential equipment, and the myriad of activities that make this hobby both enjoyable and fulfilling. Whether you are drawn to the technical intricacies of radio waves or the prospect of connecting with like-minded individuals worldwide, ham radio offers an exciting journey into a realm where exploration, learning, and community converge. So, let's embark on this adventure together and unlock the boundless potential of ham radio for beginners.

CHAPTER 1: GETTING ACQUAINTED WITH HAM RADIO

Understanding the Basics

Ham radio, also known as amateur radio, is a fascinating and dynamic hobby that allows individuals to communicate with others around the world using radio frequencies. Whether you're interested in emergency communication, community service, or just connecting with fellow enthusiasts, getting acquainted with ham radio can be a rewarding experience. Let's delve into the basics to help you understand this diverse and engaging hobby.

What is Ham Radio?

Ham radio refers to the use of designated radio frequencies for non-commercial communication by licensed operators. Unlike other forms of communication, ham radio operators (hams) are allowed to build, experiment with, and operate their own radio equipment. This provides a unique opportunity for hands-on learning and experimentation.

Licensing and Regulations

Before diving into ham radio, it's essential to obtain the necessary license from the regulatory authority in your country. The licensing process typically involves passing an examination that tests your knowledge of basic radio theory, regulations, and operating procedures. Different license classes may grant you access to varying frequencies and modes.

Equipment and Technology

Ham radio encompasses a wide range of equipment and technologies, from simple handheld transceivers to elaborate home-based setups. Understanding the basics of transmitters, receivers, antennas, and propagation will help you make informed decisions about your equipment. Experimenting with different setups and configurations is a significant part of the learning experience.

Operating Modes

Ham radio offers a variety of operating modes, each with its unique characteristics. Some of the common modes include:

Voice (SSB, FM): Traditional voice communication using single sideband (SSB) or frequency modulation (FM).

Morse Code (CW): An efficient and classic mode using coded signals to communicate.

Digital Modes: Utilizing digital signals for data transmission, including modes like PSK31, RTTY, and FT8.

Amateur Television (ATV): Transmitting and receiving video signals.

Exploring these modes allows hams to communicate in diverse ways, adapting to different conditions and preferences.

Emergency Communication and Public Service
Ham radio operators play a crucial role in emergency communication and public service. During disasters or when regular communication infrastructure fails, hams often step in to provide

essential communication links. Participating in public service events, such as marathons or parades, allows operators to hone their skills while contributing to their communities.

Community and Elmering

Elmering is a term used in the ham radio community to describe mentoring or guiding newcomers. Joining local clubs, attending hamfests (ham radio gatherings), and engaging with experienced operators can provide valuable insights and support. The ham radio community is known for its welcoming and helpful nature, making it easy for newcomers to learn and grow in the hobby.

Getting acquainted with ham radio involves a combination of technical knowledge, practical skills, and a passion for communication.

Whether you're interested in making friends around the world, participating in emergency response efforts, or simply exploring the realms of radio technology, ham radio offers a diverse and exciting journey. Obtain your license, explore different modes, connect with the community, and enjoy the endless possibilities that ham radio has to offer.

History and Evolution

Ham radio, also known as amateur radio, is a fascinating hobby that has been connecting people across the globe for over a century. Its roots can be traced back to the early experiments with wireless communication in the late 19th and early 20th centuries. Let's delve into the history and evolution of ham radio, exploring the key milestones that have shaped this unique and enduring pastime.

1. Early Experiments:

The story of ham radio begins with the pioneering work of individuals like Guglielmo Marconi, who conducted groundbreaking experiments in wireless communication during the late 1800s. These early efforts laid the foundation for the development of radio technology and inspired countless enthusiasts to explore its possibilities.

2. Amateur Radio Emerges:

As radio technology advanced, amateur radio operators, or "hams," began to experiment with and build their own radio equipment. In the early 20th century, governments started allocating specific radio frequencies for amateur use. This marked the formal beginning of amateur radio as a distinct and regulated hobby.

3. World War II and Postwar Growth:

The significance of amateur radio became even more apparent during World War II, as many skilled operators contributed to wartime communication efforts. After the war, the hobby experienced a surge in popularity, and ham radio clubs and organizations began to emerge globally.

4. Licensing and Regulations:

To ensure responsible and organized use of the radio spectrum, governments around the world introduced licensing and regulatory frameworks for amateur radio operators. These regulations specify the frequency bands allocated for amateur use, define licensing requirements, and establish operating protocols.

5. Technological Advances:

Over the decades, ham radio has evolved alongside technological advancements. The transition from analog to digital modes, the development of sophisticated transceivers, and the integration of computers into radio operations have transformed the hobby. Satellite communication and the use of space-based repeaters have added new dimensions to the capabilities of amateur radio.

6. Emergency Communication:

One of the significant contributions of amateur radio is its role in emergency communication. Hams have often played a crucial part in providing communication support during natural disasters, emergencies, and situations where traditional communication infrastructure may be compromised.

7. Community and Collaboration:

The ham radio community is known for its camaraderie and spirit of collaboration. Enthusiasts come together through local clubs, international contests, and events to share knowledge, exchange ideas, and foster friendships. The global nature of amateur radio allows operators to connect with people from diverse cultures and backgrounds.

8. Licensing Levels and Education:

Amateur radio licenses are typically tiered, with different levels indicating varying levels of technical proficiency. Licensing exams cover topics such as radio theory, regulations, and operating procedures. Many hams actively engage in continuous learning, participating in workshops and pursuing advanced certifications.

In conclusion, getting acquainted with ham radio offers a journey into a rich and dynamic history. The hobby has not only witnessed remarkable technological advancements but has also played a vital role in fostering international friendships and contributing to public service. As we navigate the digital age, ham radio continues to capture the imagination of new generations, providing a unique blend of technology, community, and exploration. Whether you are interested in the technical aspects or the social aspects of the hobby, there's a place for everyone in the world of ham radio.

Licensing Requirements

Ham radio, also known as amateur radio, is a fascinating hobby that allows individuals to explore the world of radio communication. Whether you're interested in emergency communication, making friends around the globe, or experimenting with technology, ham radio offers a diverse range of opportunities. However, before you can start transmitting on ham radio frequencies, you need to obtain a license. Let's delve into the licensing requirements for getting started with ham radio.

1. Understanding the Basics:

Ham radio operates on designated frequency bands allocated by international agreements. These bands vary in frequency and usage, from shortwave communication to satellite transmissions. To participate legally, you must adhere to the regulations set forth by your country's telecommunications authority.

2. Licensing Classes:

Licensing for ham radio operators is tiered into different classes, each granting specific privileges. The classes often include Technician, General, and Extra (or similar titles), with each subsequent class providing access to a broader range of frequencies and modes.

Technician Class: This entry-level license grants privileges on VHF (Very High Frequency) and UHF (Ultra High Frequency) bands, allowing for local communication and some limited long-distance contacts.

General Class: With additional privileges on HF (High Frequency) bands, the General Class license opens up opportunities for more extended-range communication, including international contacts.

Extra Class: The highest class provides access to all amateur radio frequencies and modes. It is a recognition of advanced knowledge and dedication to the hobby.

3. Studying for the Exam:

To obtain a ham radio license, you'll need to pass an examination administered by your country's telecommunications regulatory body. The exam typically covers radio theory, regulations, and operating practices. Numerous study materials, online courses, and practice exams are available to help you prepare.

4. Finding Exam Sessions:

Exams are conducted by volunteer examiners (VEs) who are licensed amateurs themselves. Exam sessions are held regularly, and you can find upcoming sessions in your area through local amateur radio clubs, online forums, or the telecommunications regulatory authority's website.

5. Taking the Exam:

The ham radio exam is multiple-choice and is usually divided into sections corresponding to different aspects of radio operation. Don't be intimidated; many resources are available to help you prepare. Once you pass the exam, you'll receive your license, and you can start enjoying the world of amateur radio.

6. Operating within Legal Limits:

After obtaining your license, it's crucial to operate within the legal limits defined by your license class. This includes adhering to frequency restrictions, power limitations, and any other regulations outlined in your license.

7. Continuing Education:

The learning doesn't stop after obtaining your license. Ham radio operators often engage in continuous learning to stay updated on technology advancements, operating procedures, and emergency communication protocols. Many opportunities for ongoing education and training exist within the amateur radio community.

In conclusion, obtaining a ham radio license is a rewarding step towards participating in a dynamic and global community of radio enthusiasts. By understanding the licensing requirements and embracing the learning process, you can unlock the full potential of ham radio and enjoy the countless benefits it has to offer.

CHAPTER 2: SETTING UP YOUR AMATEUR RADIO STATION

Essential Equipment

Setting up your own amateur radio station is an exciting and rewarding endeavor that allows you to explore the world of radio communication. Whether you're a seasoned ham radio operator or a beginner looking to get started, having the right equipment is crucial. In this guide, we'll cover the essential equipment you need to set up a fully functional amateur radio station.

Transceiver:

The heart of any amateur radio station is the transceiver. This device combines a transmitter and receiver in one unit, allowing you to both send and receive radio signals. When choosing a transceiver, consider factors such as frequency range, power output, and modulation modes. Popular brands include Yaesu, Icom, and Kenwood.

Antenna:

An antenna is a critical component that determines the efficiency of your radio station. The type of antenna you choose depends on your operating preferences, available space, and frequency bands you intend to use. Common types include dipole antennas, vertical antennas, and Yagi antennas. Properly installing and tuning your antenna is crucial for optimal performance.

Power Supply:

A stable power supply is essential to keep your station running smoothly. Ensure that your power supply meets the voltage requirements of your transceiver and other equipment. Many operators prefer using a linear power supply for its reliability, while others opt for switch-mode power supplies for their efficiency.

SWR Meter:

Standing Wave Ratio (SWR) meters are vital for antenna tuning. They help you adjust your antenna for optimal performance by indicating the efficiency of the power transfer between the transmitter and the antenna. A low SWR ensures that most of the transmitted power reaches the antenna, minimizing signal reflections.

Coaxial Cable:

Coaxial cables are used to connect your transceiver to the antenna. Select high-quality coaxial cables with the appropriate impedance for your setup. Lower-loss cables are preferred for longer runs, as they minimize signal attenuation. Pay attention to connectors as well, ensuring they match your equipment.

Tuner:

An antenna tuner is useful for adjusting the impedance match between your transceiver and the antenna. This is especially important when using antennas that may not be perfectly matched to the desired frequency. Automatic Antenna Tuners (ATUs) make this process more convenient, allowing for quick adjustments.

Headphones:

A good pair of headphones is crucial for clear communication, especially in noisy environments. Look for headphones with comfortable ear cups and good sound quality. Some operators also prefer models with noise-canceling features to eliminate background noise.

Microphone:

The microphone is your voice into the radio waves. Choose a microphone that is compatible with your transceiver and suits your preferences. Some operators prefer handheld microphones, while others opt for desk or boom microphones. Ensure that the microphone provides clear and crisp audio.

Setting up your amateur radio station requires careful consideration of the essential equipment mentioned above. As you build your station, take the time to research and choose equipment that suits your specific needs and preferences. With the right gear, you'll be ready to explore the exciting world of amateur radio communication. Good luck, and enjoy your journey into the realm of ham radio!

Antennas and Their Types

Setting up an amateur radio station is an exciting endeavor that allows enthusiasts to explore the world of radio communication. One crucial aspect of an amateur radio station is the selection and installation of antennas, which play a pivotal role in transmitting and receiving signals. In this guide, we'll delve into the world of antennas and explore different types to help you make informed decisions when setting up your station.

Understanding the Basics:

Before delving into antenna types, it's essential to understand some fundamental concepts. Antennas serve as the interface between your radio equipment and the electromagnetic waves propagating through the air. They convert electrical signals into radio waves for transmission and vice versa for reception.

Frequency Considerations:

Different antennas are designed for specific frequency ranges. The frequency of your radio signals will determine the type of antenna that is most effective. For instance, higher frequencies generally require smaller antennas, while lower frequencies may demand larger structures.

Antenna Types:

Dipole Antennas:

A basic and widely used type.

Consists of a straight wire, typically half the wavelength of the desired frequency.

Easy to construct and cost-effective.

Suitable for a broad range of frequencies.

Vertical Antennas:

Commonly used for mobile and base station setups.

Ideal for frequencies between 20 meters and 160 meters.

Requires a good ground plane for efficient operation.

Yagi-Uda Antennas:

Directional antennas with high gain.

Comprise multiple elements, including a driven element, reflector, and director.

Excellent for point-to-point communication over longer distances.

Loop Antennas:

Constructed in the shape of a loop, often a circle or rectangle.

Suitable for limited space installations.

Resonant loops are efficient and have low noise.

Beam Antennas:

High-gain directional antennas.

Consist of multiple elements arranged in a specific pattern.

Suitable for long-distance communication and contesting.

Wire Antennas:

Versatile and easy to install.

Can be configured as horizontal, sloping, or vertical.

Examples include the longwire and end-fed antennas.

Installation Tips:

Height Matters:

Mount your antenna as high as possible for improved signal propagation.

Avoid obstacles like buildings and trees that can obstruct the signal path.

Balancing Act:

Achieve a good balance between gain and directionality based on your communication needs.

Consider the radiation pattern of the antenna.

Grounding:

Properly ground your antenna to enhance safety and performance.

Use suitable grounding materials and follow recommended practices.

SWR Tuning:

Adjust the Standing Wave Ratio (SWR) to optimize your antenna's performance.

Use an SWR meter to ensure efficient power transfer.

Legal Considerations:

Adhere to local regulations and restrictions on antenna height and placement.

Consult with local authorities if necessary.

Selecting and setting up the right antenna is crucial for the success of your amateur radio station. Consider your communication goals, available space, and budget when choosing an antenna type. Experimenting with different configurations can also enhance your understanding of radio wave propagation and improve your overall station performance. As you embark on this journey, remember that the world of amateur radio is vast, and the learning experience is as rewarding as the communication itself.

Creating a Safe and Efficient Operating Space

Amateur radio, also known as ham radio, is a rewarding hobby that allows enthusiasts to communicate with others around the world using radio frequencies. Whether you are a newcomer to the hobby or a seasoned operator, setting up a safe and efficient operating space is crucial for an enjoyable and successful experience. Here's a guide to help you create an ideal environment for your amateur radio station.

1. Selecting a Suitable Location:

Choose a location with minimal RF interference, away from electronic appliances and power lines. If possible, opt for a dedicated room to establish your station. Ensure that the space has proper ventilation and is comfortable for long operating sessions.

2. Equipment Layout and Organization:

Arrange your equipment thoughtfully to maximize efficiency and minimize clutter. Keep essential items within easy reach and organize cables neatly to avoid tangling. Consider using cable organizers and racks to maintain a clean and tidy workspace.

3. Proper Grounding:

Grounding is crucial for safety and optimal performance. Ground all equipment, antennas, and accessories according to your local regulations. Ensure that your station is properly bonded to minimize the risk of electrical shock and to protect against lightning strikes.

4. Power Supply Considerations:

Invest in a reliable and clean power source to avoid interference and equipment damage. Use surge protectors and uninterruptible power supplies (UPS) to safeguard your gear from power fluctuations and outages.

5. Lighting:

Adequate lighting is essential for comfortable operation, especially during nighttime sessions. Use adjustable lighting to reduce glare on your equipment and minimize eye strain. Consider using LED lights to save energy and minimize heat generation.

6. Ergonomics:

Design your operating position with ergonomics in mind. Choose a comfortable chair and position your equipment at the right height and angle to reduce fatigue during extended operating sessions. Consider anti-fatigue mats for standing stations.

7. Noise Reduction:

Identify and eliminate sources of ambient noise, as it can affect the clarity of your communication. Use ferrite chokes on cables, and consider acoustic treatments for your operating space to reduce echoes and unwanted background noise.

8. Ventilation and Cooling:

Amateur radio equipment can generate heat, especially during prolonged use. Ensure proper ventilation to prevent overheating. Consider installing fans or air conditioning, especially in confined spaces, to maintain a comfortable operating temperature.

9. Safety Measures:

Familiarize yourself with emergency procedures and keep a first aid kit nearby. Ensure that fire extinguishers are easily accessible. Clearly mark emergency exits and have a communication plan in place with family members or cohabitants.

Setting up an amateur radio station involves careful planning to create a safe and efficient operating space. By selecting the right location, organizing equipment thoughtfully, ensuring proper grounding, and implementing ergonomic design, you can enhance your overall amateur radio experience. Remember to stay informed about local regulations and safety practices to enjoy your hobby responsibly.

CHAPTER 3: NAVIGATING THE FREQUENCIES

Bands and Modes Overview

Navigating the frequencies in the vast realm of radio communication involves understanding the various bands and modes available. Radio frequencies are divided into different bands, each serving specific purposes and accommodating diverse communication needs. Additionally, modes define how information is transmitted over these frequencies. This article provides an overview of radio frequency bands and communication modes, offering insights into the fascinating world of wireless communication.

Radio Frequency Bands:

Very Low Frequency (VLF):

Frequency Range: 3 kHz - 30 kHz

Application: Navigation, submarine communication

Low Frequency (LF):

Frequency Range: 30 kHz - 300 kHz

Application: Maritime navigation, time signals

Medium Frequency (MF):

Frequency Range: 300 kHz - 3 MHz

Application: AM broadcasting, aviation communication

High Frequency (HF):

Frequency Range: 3 MHz - 30 MHz

Application: Shortwave broadcasting, amateur radio, aviation, maritime communication

Very High Frequency (VHF):

Frequency Range: 30 MHz - 300 MHz

Application: FM broadcasting, air traffic control, land mobile communication

Ultra High Frequency (UHF):

Frequency Range: 300 MHz - 3 GHz

Application: TV broadcasting, satellite communication, military communication

Super High Frequency (SHF):

Frequency Range: 3 GHz - 30 GHz

Application: Radar, satellite communication, microwave ovens

Extremely High Frequency (EHF):

Frequency Range: 30 GHz - 300 GHz

Application: Radio astronomy, millimeter-wave communication

Communication Modes:

Amplitude Modulation (AM):

Modulates the amplitude of the carrier signal.

Common in broadcast radio and aviation communication.

Frequency Modulation (FM):

Modulates the frequency of the carrier signal.

Used in FM radio broadcasting, two-way radios, and air traffic control.

Single Sideband (SSB):

Suppresses one of the sidebands and carrier to save bandwidth.

Popular in amateur radio and long-distance communication.

Continuous Wave (CW):

Utilizes a continuous wave for Morse code communication.

Widely used in amateur radio and maritime communication.

Phase Modulation (PM):

Modulates the phase of the carrier signal.

Common in satellite communication and digital radio systems.

Digital Modes:

Utilize digital encoding for information transmission.

Examples include Morse code, PSK31, and FT8.

Understanding the characteristics of each band and mode is crucial for effective communication.

Factors such as propagation characteristics, atmospheric conditions, and regulatory constraints influence the choice of bands and modes for specific applications.

Navigating the Frequencies:

Frequency Allocation:

Understand the regulatory allocations for different frequency bands in your region.

Propagation Conditions:

Be aware of how atmospheric conditions affect signal propagation in different bands.

Licensing Requirements:

Obtain the necessary licenses for operating in certain frequency bands, especially in amateur radio and commercial communication.

Equipment Selection:

Choose the appropriate radio equipment based on the desired frequency band and communication mode.

Antenna Considerations:

Different frequencies require different antenna designs. Choose antennas that match the frequency band for optimal performance.

Interference Management:

Be mindful of potential interference from other radio signals and electronic devices.

Stay Informed:

Keep abreast of technological advancements, regulatory changes, and best practices in radio communication.

Navigating the frequencies requires a combination of technical knowledge, regulatory compliance, and practical experience. Whether you are a radio enthusiast, a professional in the field, or simply curious about wireless communication, understanding the intricacies of bands and modes enhances your ability to explore and communicate effectively in the diverse world of radio frequencies.

Tuning In: Scanning and Listening

In a world saturated with invisible waves and frequencies, the art of navigating through them has become increasingly important. From radio signals to Wi-Fi networks, understanding how to tune in, scan, and listen to frequencies can provide a wealth of information and opportunities. In this exploration, we delve into the intricacies of navigating the frequencies, unlocking a world that is often unheard and unseen.

The Spectrum of Frequencies:

The electromagnetic spectrum encompasses a vast array of frequencies, ranging from radio waves with long wavelengths to gamma rays with short wavelengths. Each frequency band serves a unique purpose, offering a diverse range of applications. Navigating this spectrum allows us to tap into various technologies and communications systems that define the modern era.

Tuning In: The Basics:

Tuning in involves selecting a specific frequency to access information or communication signals. This concept is most commonly associated with radio, where listeners tune in to different stations by adjusting the frequency on their devices. In the digital age, tuning in can also refer to accessing specific channels on television or connecting to wireless networks.

Scanning the Airwaves:

Scanning is the process of systematically searching for signals across a range of frequencies. This technique is commonly used in radio communication and is essential for finding available channels or identifying potential interference. Radio scanners, for example, can sweep through a wide range of frequencies, enabling users to discover and monitor various transmissions.

Listening Beyond the Obvious:

Beyond traditional communication systems, there are frequencies that exist in realms we may not consider. The natural world, for instance, is filled with sounds and frequencies that animals use for communication and navigation. Exploring these frequencies opens up opportunities for understanding ecosystems and the intricate connections between species.

Wireless Technologies and Connectivity:

In the digital age, wireless technologies dominate our lives. Understanding how to navigate the frequencies is crucial for accessing Wi-Fi networks, Bluetooth devices, and other wireless communication systems. As we rely more on interconnected devices, the ability to tune in and connect seamlessly becomes increasingly valuable.

Challenges and Ethical Considerations:

While navigating frequencies offers numerous benefits, it also raises ethical concerns. Privacy issues arise when considering the unintentional interception of communication signals. Striking a balance between accessing information and respecting privacy is a challenge that comes with the ability to navigate frequencies effectively.

The Future of Frequency Navigation:

Advancements in technology continue to push the boundaries of frequency navigation. From the exploration of extraterrestrial frequencies to the development of advanced communication systems, the future promises new opportunities and challenges. As we progress, it becomes essential to cultivate a responsible and ethical approach to navigating the frequencies.

Navigating the frequencies is a skill that empowers us to connect, communicate, and explore the unseen realms of our world. Whether tuning in to a favorite radio station, scanning for available Wi-Fi networks, or listening to the natural frequencies of the environment, understanding the intricacies of frequency navigation opens doors to a world filled with information and possibilities. As we embrace this skill, let us do so responsibly, respecting the privacy and ethical considerations that come with the ability to navigate the frequencies.

Joining Conversations and Making Contacts

In the vast landscape of human interaction, communication is the key that unlocks doors and connects individuals across diverse frequencies. Navigating these frequencies involves the art of joining conversations and making meaningful contacts, transcending barriers and fostering genuine connections. Whether in personal or professional settings, the ability to navigate frequencies can greatly enhance one's social and networking skills.

Understanding the Frequencies

Frequencies in this context refer to the various channels of communication, both verbal and non-verbal, that people engage in. These can include face-to-face conversations, virtual discussions, written correspondence, and even body language. Each frequency carries its own nuances, and adept navigators can seamlessly transition between them to establish rapport and build relationships.

Active Listening as a Navigational Tool

A crucial aspect of navigating frequencies is active listening. Actively engaging in a conversation requires more than just hearing words; it involves understanding the underlying emotions, intentions, and perspectives of others. By tuning in to the subtleties of communication, one can respond appropriately and contribute meaningfully to ongoing discussions.

Entering Conversations with Confidence

Joining conversations can be intimidating, especially in unfamiliar or professional settings. Confidence plays a vital role in this process. Approach conversations with a positive mindset, maintain eye contact, and be mindful of your body language. Confidence not only makes you more approachable but also helps you convey your ideas effectively.

Adapting to Virtual Frequencies

In today's digital age, navigating virtual frequencies is a crucial skill. Whether through emails, video calls, or social media platforms, the ability to communicate effectively online is essential. Pay attention to the tone of your written communication, utilize video calls for a more personal touch, and stay mindful of time zones and cultural differences.

Building Contacts with Authenticity

Making contacts goes beyond collecting business cards or LinkedIn connections. Authenticity is the cornerstone of meaningful relationships. Share your thoughts genuinely, express interest in others, and be open to learning from diverse perspectives. Authentic connections often lead to long-lasting relationships that extend beyond immediate needs or transactions.

Networking as a Two-Way Street

Networking is a powerful tool for expanding your contacts and opportunities. However, it's important to approach it as a two-way street. Contribute value to others, offer assistance, and be genuinely interested in their success. This reciprocity strengthens connections and transforms networking into a mutually beneficial endeavor.

Navigating Across Cultures

In our interconnected world, navigating frequencies often involves engaging with individuals from diverse cultural backgrounds. Cultural sensitivity is crucial in these interactions. Learn about different customs, traditions, and communication styles to foster understanding and build bridges across cultural divides.

Continuous Learning and Adaptation

The art of navigating frequencies is dynamic, requiring continuous learning and adaptation. Stay updated on communication trends, be open to feedback, and refine your skills based on experiences. Flexibility and a willingness to learn will help you stay attuned to evolving social dynamics.

Navigating the frequencies of human interaction is a lifelong journey filled with opportunities for growth and connection. By mastering the art of joining conversations and making contacts, individuals can enrich their personal and professional lives, fostering a network of relationships that contribute to personal success and the well-being of the community at large.

CHAPTER 4: OPERATING PROCEDURES AND ETIQUETTE

Standard Protocols

Operating procedures and etiquette play a crucial role in maintaining a well-organized and efficient work environment. Standard protocols are established guidelines that ensure consistency, safety, and professionalism in various settings. Whether in a corporate office, healthcare facility, or any other workplace, adherence to these protocols is essential for the smooth functioning of operations.

I. Importance of Operating Procedures:

Consistency: Standard protocols create a consistent framework for performing tasks. This consistency helps streamline processes and reduces the likelihood of errors.

Efficiency: Well-defined operating procedures enhance efficiency by providing a structured approach to completing tasks. Employees can follow a set path, reducing time and effort spent on decision-making.

Safety: Many protocols are designed with safety in mind. Following established procedures helps minimize the risk of accidents and ensures a secure working environment for all.

Compliance: Operating procedures often align with legal and industry regulations. Adhering to these protocols ensures that the organization stays compliant with relevant laws and standards.

II. Components of Standard Protocols:

Documentation: Clearly written and accessible documentation is fundamental. Manuals, guidelines, and procedures should be regularly updated and readily available to all team members.

Training: Proper training on established protocols is essential. This ensures that employees are aware of the correct procedures and understand the rationale behind them.

Communication: Effective communication is a key component. Protocols should be communicated clearly to all relevant parties, and any updates or changes should be promptly disseminated.

III. Etiquette in the Workplace:

Professionalism: Maintaining a professional demeanor is crucial. This includes dressing appropriately, using proper language, and conducting oneself with respect towards colleagues and clients.

Communication Etiquette: Clear and respectful communication is vital in any workplace. This involves active listening, avoiding interrupting others, and using appropriate channels for communication.

Punctuality: Being on time for meetings, deadlines, and daily tasks demonstrates reliability and respect for others' time.

Punctuality contributes to a smoothly running operation.

Collaboration: Effective teamwork is built on mutual respect and collaboration. Encouraging an open exchange of ideas and valuing the contributions of each team member fosters a positive working environment.

IV. Adapting to Change:

Flexibility: Standard protocols should not be rigid. Flexibility is key to adapting to changes in technology, industry trends, or organizational structures.

Continuous Improvement: Regularly reviewing and updating protocols allows for continuous improvement. This ensures that the organization remains adaptable and responsive to evolving circumstances.

Operating procedures and etiquette are the foundation of a successful and harmonious work environment. By adhering to standard protocols and cultivating a culture of professionalism and respect, organizations can enhance efficiency, ensure safety, and foster a positive workplace culture. Regularly reviewing and updating these protocols is essential to staying relevant and responsive in a dynamic business landscape.

Using Q Codes and Abbreviations

Amateur radio, commonly known as ham radio, is a fascinating hobby that enables individuals to communicate globally using radio frequencies. As with any form of communication, there are specific operating procedures and etiquette that help maintain efficient and respectful communication among ham radio operators. One essential aspect of this communication protocol is the use of Q codes and abbreviations.

Understanding Q Codes:

Q codes are a set of three-letter codes that were initially developed for maritime communication and later adopted by ham radio operators. These codes serve as a shorthand way of conveying information, reducing the time required to exchange messages and enhancing overall communication efficiency.

QSO (QSO?): The QSO code is used to initiate a conversation, asking if the frequency is in use. For example, "Is this frequency in use? QSO?"

QSL (QSL?): QSL is a code used to confirm receipt of a message. When a ham radio operator wants confirmation that their message has been received, they might say, "Did you receive my transmission? QSL?"

QTH (QTH?): The QTH code inquires about the station's location. For example, "Can you tell me your location? QTH?"

QRP and QRO: QRP signifies low power, while QRO indicates high power. These codes are often used to convey the transmitting power of a station. "I am running QRP" means the operator is using low power.

Efficient Use of Abbreviations:
Apart from Q codes, ham radio operators frequently employ abbreviations to streamline communication. These abbreviations help convey information concisely and are especially useful when conditions are challenging.

73 and 88: These are common sign-offs in ham radio. "73" is used as a farewell or sign-off, and "88" is often used to convey love and kisses.

CQ (CQ, CQ, CQ): CQ is a general call inviting any station to respond. Repeating it three times is common practice. "CQ, CQ, CQ, this is [callsign], calling any station."

OM (Old Man) and YL (Young Lady): These abbreviations are used to refer to male and female operators, respectively.

WX (Weather): When discussing weather conditions, ham radio operators often use "WX" as an abbreviation.

Etiquette in Ham Radio:

Listen Before Transmitting: Always listen to ongoing conversations on the frequency before transmitting to avoid unnecessary interference.

Wait for a Pause: When joining an ongoing conversation, wait for a suitable break before interjecting.

Identify Yourself Clearly: State your callsign clearly and concisely. Repeat it if necessary to ensure accurate reception.

Be Courteous: Maintain a polite and respectful tone during interactions. Avoid unnecessary disruptions or contentious discussions.

In conclusion, mastering operating procedures and etiquette in ham radio, including the use of Q codes and abbreviations, enhances communication efficiency and fosters a positive community experience. By adhering to these guidelines, ham radio operators contribute to a vibrant and respectful global network of communication.

Handling Emergency Situations

Ham radio, or amateur radio, is a fascinating hobby that allows individuals to communicate over the airwaves using designated radio frequencies. While ham radio is primarily a hobby for enthusiasts, it plays a crucial role in emergency communication during disasters or other critical situations. Operating procedures and etiquette are fundamental aspects of ham radio, especially when it comes to handling emergency situations. This guide outlines essential practices to ensure effective communication and cooperation during emergencies.

Follow Established Protocols:

In emergency situations, adherence to established protocols is crucial. Familiarize yourself with emergency procedures outlined by your local ham radio club, ARES (Amateur Radio Emergency Service), or other relevant organizations. Clear protocols streamline communication and help maintain order during crises.

Frequency Discipline:

During emergencies, specific frequencies are allocated for priority communication. It is essential to respect these allocations and refrain from using emergency frequencies for non-critical communication. This discipline ensures that emergency responders and key operators can effectively coordinate their efforts.

Emergency Nets:

Emergency nets are organized networks of ham radio operators who provide essential communication support during crises. Participate in local emergency nets, and be prepared to relay critical information as needed. Familiarize yourself with net procedures and be ready to follow the net control operator's instructions.

Practice Clear and Concise Communication:
Clarity is paramount during emergency communication. Practice clear and concise messaging, avoiding unnecessary jargon or ambiguity. Use standardized communication protocols, such as the Q-code, to convey information efficiently.

Priority Traffic:
Emergency situations may involve priority traffic, which consists of urgent messages related

to the crisis. Be aware of and respect the priority traffic system, ensuring that crucial information is prioritized and transmitted promptly.

Collaborate with Emergency Services:
Ham radio operators often work in collaboration with local emergency services. Establish contacts with relevant authorities, participate in emergency drills, and understand how ham radio can complement professional emergency communication systems.

Power Management:
In emergency situations, power resources may be limited. Efficiently manage your power supply to ensure continuous operation. Have backup power sources such as generators or batteries, and use power-saving modes when appropriate.

Respect for Others:

Maintain a respectful and cooperative attitude towards fellow operators and emergency responders. Cooperation and a positive atmosphere contribute to effective communication and overall success in managing emergencies.

Continuous Training:

Regularly participate in training exercises and drills focused on emergency communication. This helps hone your skills, familiarize yourself with new technologies, and ensures that you are well-prepared to handle different types of emergencies.

Update Emergency Contact Information:

Keep your contact information updated with relevant emergency services and organizations. This ensures that you can be reached promptly in case your assistance is required.

Operating procedures and etiquette in ham radio are essential components when it comes to handling emergency situations. By following established protocols, practicing clear communication, and collaborating with emergency services, ham radio operators play a vital role in ensuring effective communication during critical times. Continuous training and a commitment to excellence will further enhance the capabilities of ham radio operators in emergency scenarios.

CHAPTER 5: EXPLORING ADVANCED FEATURES AND TECHNOLOGIES

Digital Modes and Software

Amateur radio, or ham radio, has come a long way since its inception, evolving with technological advancements to offer enthusiasts a plethora of advanced features. One significant stride in this evolution is the integration of digital modes and software, revolutionizing the way radio communication is conducted. In this exploration, we delve into the cutting-edge world of ham radio, focusing on the advanced features and technologies that digital modes and software bring to the table.

Digital Modes: The Shift from Analog to Digital:

The traditional analog communication in ham radio has now been complemented, and in some cases replaced, by digital modes. Digital modes utilize encoding and decoding techniques to transmit data, providing advantages such as improved signal quality, error correction, and efficient use of bandwidth. Popular digital modes include:

FT8 and FT4: These modes are widely used for weak signal communication, allowing operators to make contacts under challenging propagation conditions.

PSK31: A popular phase-shift keying mode for keyboard-to-keyboard communication, ideal for low-power and narrow bandwidth operation.

JT65 and JT9: Designed for extreme weak signal conditions, these modes enable reliable communication even when signals are barely audible.

Software-Defined Radios (SDR): Transforming Radio Communication:

SDR technology has revolutionized ham radio by shifting much of the signal processing from traditional hardware components to software. This flexibility allows for a wide range of advanced features, including:

Wide Frequency Range: SDRs cover a broader frequency spectrum, enabling operators to explore various bands without the need for multiple dedicated radios.

DSP Processing: Digital Signal Processing (DSP) capabilities provide advanced filtering, noise reduction, and signal enhancement, improving overall reception quality.

Software-Based Modulation and Demodulation: SDRs can adapt to different modulation schemes through software updates, making them versatile for emerging digital modes.

Digital Voice Modes: Enhancing Clarity and Efficiency:
Digital voice modes leverage data compression and error correction techniques to transmit voice signals in a digital format. This not only

improves audio quality but also allows for more efficient use of bandwidth. Examples include:

D-STAR (Digital Smart Technologies for Amateur Radio): This digital voice and data protocol offers clear and reliable communication with features like call routing and text messaging.

DMR (Digital Mobile Radio): Widely adopted in both amateur and commercial radio, DMR provides efficient voice communication along with data capabilities.

Fusion (C4FM): Developed by Yaesu, Fusion combines voice and data transmission, offering advanced features like group communication and GPS positioning.

Logging and Station Automation Software:

Modern ham radio operators benefit from feature-rich logging software that helps manage contacts, awards, and station activities. Automation software allows for remote control and monitoring of radio equipment. Key features include:

Automatic Log Entry: Software automatically records details of each contact, simplifying the process of confirming awards and participating in contests.

Integration with Digital Modes: Logging software often integrates seamlessly with digital mode programs, streamlining the logging process during digital communications.

Remote Operation: Remote control software enables operators to control their stations from anywhere in the world, opening up new possibilities for communication and experimentation.

As ham radio continues to embrace digital modes and software, enthusiasts find themselves equipped with a powerful array of tools to enhance their communication experiences. Whether engaging in weak signal contacts, experimenting with digital voice modes, or exploring the capabilities of software-defined radios, the advanced features discussed here represent the forefront of ham radio technology. Embracing these innovations not only expands the horizons of communication but also ensures that ham radio remains a dynamic and evolving hobby for years to come.

Satellite Communications

Ham radio enthusiasts have always been at the forefront of adopting and integrating cutting-edge technologies into their hobby. One such area that has seen significant advancements is satellite communications. In recent years, ham radio operators have embraced advanced features and technologies to enhance their communication capabilities via satellites. This article delves into the exciting world of ham radio satellite communications, exploring the latest innovations and how they are transforming the amateur radio landscape.

Satellite Tracking and Antenna Systems:

Modern ham radio satellite communications rely on sophisticated tracking systems and antenna arrays. Software-defined radios (SDRs) coupled with computer-controlled rotators enable operators to accurately track and communicate with satellites as they orbit the Earth.

High-gain antennas, such as Yagi-Uda or helical antennas, have become standard, providing the directional sensitivity needed for reliable communication with satellites.

Digital Signal Processing (DSP):

DSP plays a crucial role in extracting weak signals from the noise inherent in satellite communications. Advanced DSP algorithms are employed to enhance signal quality and mitigate interference, enabling clear and reliable communication even in challenging conditions.

Transponder Technologies:

Many ham radio satellites utilize transponders, which are devices that receive signals on one frequency and retransmit them on another. Advances in transponder technology have led to higher data rates, improved bandwidth efficiency, and increased flexibility for different communication modes.

Software-Defined Radios (SDRs):

SDRs have revolutionized ham radio by providing a flexible and programmable platform. In satellite communications, SDRs allow operators to adapt to different satellite transponder configurations, experiment with modulation schemes, and implement real-time signal processing.

CubeSats and NanoSats:

The proliferation of small satellites, such as CubeSats and NanoSats, has expanded opportunities for ham radio satellite communication. These small satellites often carry amateur radio payloads, providing a cost-effective means for hams to engage in space communications.

Integrated Ground Stations:

Ham radio operators are building integrated ground stations that combine satellite tracking, antenna control, and communication equipment into a seamless system. This integration streamlines the process of connecting with satellites and enhances the overall user experience.

Global Positioning System (GPS) Integration:

GPS technology is integrated into many modern satellite communication setups. This integration allows for automatic tracking and pointing of antennas, making it easier for operators to stay aligned with satellites as they move across the sky.

Online Resources and Communities:

The internet has facilitated the sharing of satellite pass predictions, propagation data, and real-time tracking information. Online communities and forums provide a platform for operators to exchange experiences, troubleshoot issues, and collaborate on satellite-related projects.

Ham radio satellite communications have evolved significantly with the integration of advanced features and technologies. From precise tracking systems and DSP algorithms to SDRs and CubeSats, enthusiasts are pushing the boundaries of what's possible in amateur radio. These innovations not only enhance the technical aspects of the hobby but also foster a sense of community as operators collaborate to explore the vast possibilities of ham radio in the satellite realm. As technology continues to advance, ham radio enthusiasts can look forward to even more exciting developments in satellite communications.

Building Simple Projects: A Practical Approach

Ham radio, or amateur radio, has come a long way since its inception, evolving with advanced features and cutting-edge technologies. This fascinating hobby not only connects enthusiasts worldwide but also offers a platform for experimentation and innovation. In this exploration, we delve into the advanced features and technologies that have transformed ham radio, with a focus on building simple yet practical projects.

Software-Defined Radios (SDRs):

Modern ham radios often utilize Software-Defined Radios, marking a significant departure from traditional hardware-based radios. SDRs enable users to reconfigure their radios through software, opening up a world of possibilities. Explore projects that involve building and customizing your own SDR, allowing you to experiment with different modulation schemes and signal processing techniques.

Digital Modes and Signal Processing:
Ham radio has embraced digital modes for more efficient communication. Projects involving digital signal processing (DSP) can be both educational and rewarding. Learn to implement digital modes such as PSK31, FT8, or JT65, and build interfaces to connect your radio to a computer for decoding and encoding digital signals.

Automatic Packet Reporting System (APRS):

APRS integrates ham radio with GPS technology, enabling real-time tracking and data exchange. Building a simple APRS project can involve setting up a tracking device in your car or creating a weather station that transmits data over the airwaves. Learn how to interface a GPS module with your radio and experiment with different data formats for APRS transmissions.

Satellite Communication:

Ham radio operators can now communicate through satellites orbiting the Earth, opening up a whole new dimension to the hobby. Explore building simple satellite communication projects, including tracking satellite passes and designing antennas for satellite communication. Learn the basics of Doppler shift compensation and frequency coordination for successful satellite contacts.

High-Frequency (HF) Antenna Tuners:
Building a high-frequency antenna tuner can be a practical project for ham radio operators interested in maximizing their HF station's performance. Learn about impedance matching, variable capacitors, and inductors to create a tuner that adapts your antenna system to the changing conditions of the HF bands.

Ham radio enthusiasts can embark on a journey of discovery by exploring advanced features and technologies in their projects. Whether it's experimenting with SDRs, diving into digital signal processing, incorporating APRS for data exchange, venturing into satellite communication, or enhancing HF station performance with antenna tuners, the possibilities are vast. By taking a practical approach to these projects, hams can not only deepen their understanding of the technology but also contribute to the ever-evolving landscape of amateur radio.

CONCLUSION

In conclusion, "Ham Radio for Beginners: The Ultimate Manual Guide to Build and Operate Your Amateur Radio as a Newbie" stands as an invaluable resource that not only demystifies the intricate world of amateur radio but also empowers beginners with the knowledge and skills needed to embark on an exciting journey into this unique hobby. The comprehensive nature of the book ensures that readers, regardless of their prior experience, can grasp the fundamentals of ham radio operation and gradually progress to more advanced techniques.

One of the notable strengths of this manual lies in its careful balance between theory and practical application. The author has adeptly broken down complex technical concepts into digestible portions, allowing novices to build a solid foundation before delving into the hands-on aspects of setting up and operating their amateur radio stations. The step-by-step instructions, accompanied by clear illustrations and explanations, enhance the learning experience, making it accessible to a wide audience.

Moreover, the book not only imparts technical knowledge but also instills a sense of community and camaraderie that is integral to the world of ham radio. By emphasizing the importance of licensing, etiquette, and respectful communication, the author promotes a culture of responsible and ethical operation. This not only ensures a positive experience for beginners but also contributes to the overall well-being of the amateur radio community.

As the title suggests, the guide is truly the ultimate manual for beginners, covering a spectrum of topics ranging from selecting the right equipment to understanding different modes of communication. The author's passion for the subject matter shines through, creating an engaging and enjoyable reading experience. The anecdotes, real-life examples, and practical tips peppered throughout the book add a personal touch, bridging the gap between theoretical knowledge and practical application.

Beyond the technical aspects, the book serves as a motivational tool, encouraging newcomers to explore the vast possibilities within the ham radio hobby. It sparks curiosity and fosters a sense of adventure, challenging readers to experiment, innovate, and contribute to the ever-evolving landscape of amateur radio.

In a world dominated by ever-advancing technology, "Ham Radio for Beginners" not only preserves the rich tradition of amateur radio but also positions it as a relevant and rewarding pursuit in the modern age. The manual is a testament to the enduring appeal of ham radio, showcasing its potential to bring people together, foster learning, and create lasting connections that transcend geographical boundaries.

In conclusion, this comprehensive guide is a must-read for anyone seeking to embark on a fulfilling journey into the realm of amateur radio. Whether you are a complete novice or have some prior experience, "Ham Radio for Beginners" provides the perfect roadmap to navigate the exciting world of ham radio, ensuring that each reader emerges with the skills, knowledge, and enthusiasm to thrive in this dynamic and ever-evolving hobby.